FISSION
Population

Overpopulation and Destruction
of
Natural Resources and Ecological Balance
of the Planet Earth

P. Clay Sherrod
Arkansas Sky Observatories

FISSION POPULATION

Overpopulation and Destruction
of
Natural Resources and Ecological Balance
of the Planet Earth

by

P. Clay Sherrod
Arkansas Sky Observatories

Revision 1

Arkansas Sky Observatories Publications
Available from: http://arksky.org/publications
Also available at Amazon, Barnes & Noble and other
booksellers
ISBN 978-0-359-60492-0

Introduction

It is somewhat oxymoronic that in the past several decades, for the first time in human history, the number of births *per capita* have decreased. This would seem like good news in terms of over population of the world we live on, but when you consider the exact *"capita"* on which the per capita is based, it appears that we have gained so little, if anything, in slowing the destruction of planet Earth's natural ecosystem that has provided all of the billions of species of biology – including that of mankind – with the abundance of both space and resources for what at one time must have seemed infinite.

Of all life on Earth, only humans can control their destiny; they have been given that ability, that right, because of the nature of our intellect which governs logic, processing of thought, and creativity. We are not the only creatures who can build a house – a bird, beaver, and insects can do that along with so many other living things. But we can go beyond the lodging for survival – we can stretch our creativity unlike other species to change the environment so that it suits our needs; if it is too warm we create air conditioning, if we do not like being rained on, we invent the umbrella.

It is that ability that sets humans apart from the rest of the living world. Further to that, the human specie, *homo sapiens*, is part of the animal kingdom and enjoys the ability to move and not be planted, like most of the biologically active portion of the plant kingdom. Nonetheless, we all are living, and we are all creatures of a planet....a very small planet and certainly one that is NOT infinite.

Now that we reach the 21st century we are experiencing just how finite our globe it. No longer is it flat and without boundaries; it is a sphere that has accumulated its resources and a living habitat for the evolution of biology over the past 3 billion years. The overall cosmic design of Earth was as near a perfect and ordering mechanism as could ever be imagined.

All species of the planet take from the Earth as they live the spans of their lives; the food, water, gases as resources as necessary; but the natural recycling ability of our planet from a geological, meteorological and biological viewpoint was nearly perfect in such a way that through the death of species which consume, their bodies also provided back into the ecosystem the resources needed for the cycle to continue.

Only with the introduction of the most intellectually capable living thing that the world had ever witnessed some three million years ago, did we see an entity on Earth that did NOT give back equal to what it consumed. For millions of years *homo sapiens* co-existed equally among the life forms from the standpoint of non-destruction.

It was when the ability and creativity evolved to a level where mankind became a selfish creature that the remarkable recycling ecosystem of this planet ceased to function like a perpetual motion machine. Exponentially resources broke down because a return to the earth was not there; and then followed

the explosion of population of these homo sapiens -
the Fission Population – which may ultimately lead
to self-imposed extinction.

For the moment imagine that you are a crewman aboard a deep water submarine lurking somewhere near the Arctic circle. Your mind might be on the unseen enemy, should one be about above or below the frozen waters above you. Year after year, you are watching….waiting. But you see no enemy ships, no threatening torpedoes, no foreign intruders.

There IS something, however, you do notice; a potential devastating threat for humanity that seems much less threatening and more sublime than machines of war.

Ice and the Survival of the Human Race

Yale University professor *Peter Wadhams* ("A Farewell to Ice", 2016) was an oceanographic scientist studying the thickness of the north polar ice sheet while aboard a British nuclear submarine in the early 1970s. Each year that he joined the crew aboard the vessel, he was noticing something peculiar and in his own words, "*….a stunning decline in the sea ice covering the northern polar regions. A more than 50 percent drop in the extent in summer and even steeper reduction in ice volume.*"

Just a few decades ago, ice 10 to 12 feet thick covered the North Pole, with sub-surface ice ridges in some parts of the Arctic extending down to 150 feet. Now, that ice is long gone, while the total volume of Arctic sea ice in late summer has declined, according to two estimates, by 75 percent in half a century.

In 2015, Wadhams recorded that the ice shelf over the Arctic Ocean was nearly the lowest ever on record, and he speculates that there is coming an "ice-free" North Pole during warm months, this possibly beginning as early as 2020. If it continues, he believes that there will be four months of no ice within the Arctic Circle within the decade and rapidly expanding to five months.

Having no ice covering the Arctic Ocean seems nearly impossible for most of us, and few have any idea the implications of just how serious this can be.

"As ocean and air temperatures in the Arctic rise, this adds more water vapor to the atmosphere, since warmer air holds more moisture. Water vapor is itself a greenhouse gas, trapping outgoing long-wave radiation and holding heat closer to the surface of the earth. With air temperatures rising in many parts of the Arctic by several degrees F in recent decades, water vapor concentration has gone up by more than 20 percent, adding to Arctic warming," explains Wadhams.

Both polar caps have been dubbed "The Air Conditioning of Mother Earth", these reflectively mirror-like ice caps helping to regulate the entire climate of Earth, and stabilize predictable – and livable – climate for periods of thousands of years.

In fact, it is noted that the once pearly-white brilliant reflective ice cap at the North Pole is now turning blue....not from contaminants, but from the sheer lack of any depth whatsoever to the sheet of ice that once covered the vast waters of the Arctic Circle.

The declining glacial ice of Greenland is now the largest single contributor to global sea level rise, its melting ice cap adding more that 72 cubic miles of water to the ocean each year. This would mean a sea level rise by the end of the 21st century of two to three feet minimum according to the Intergovernmental Panel on Climate Change (IPCC) .

But even this grim forecast is being revised upwards, with serious implications for the survival of low-lying cities like Miami, New Orleans, London, Venice, and Shanghai, or of potentially below sea level coastlines like Bangladesh. These major cities, and many others are very vulnerable to the rising of the ocean levels; consider the populations (2017 or 2018 whichever data is most current):

City - Population

Miami – 470,000
New Orleans – 395,000
London – 8,630,000
Venice – 262,000
Shanghai – 24,100,000
Bangladesh – 164,000,000

….or roughly about 200 million people now living in what could be under water by 2050.

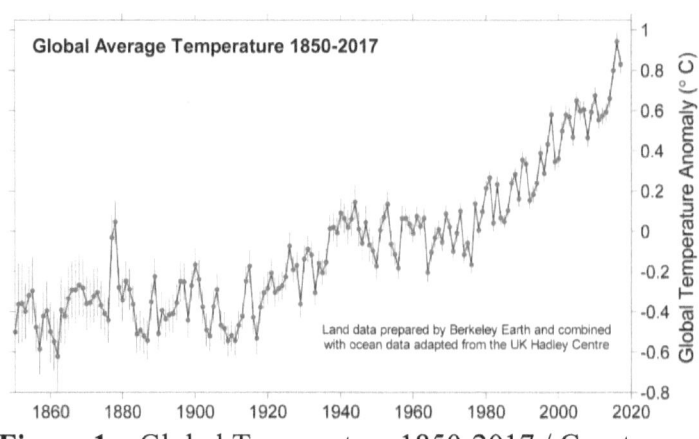

Figure 1 – Global Temperature 1850-2017 / Courtesy Berkeley (CA) Earth

*   *   *

Noting the map in the following Figure 2, the ice coverage in the Arctic hit record lows through the winter of 2017. In March, when the sea ice should reach its largest area extent of each year, for 2017 it was lower than it ever had been in the history of satellite recording, over 40 years.   This ice typically begins to melt in spring and by 2017 the melt began a month earlier than normal The Bering and Chukchi Seas along Alaska's coast remained ice-free longer into the fall than ever before recorded.

By November, 2017, NASA reported that two to four times as many coastal glaciers around Greenland are at risk of accelerated melting. Glacial coverage in Greenland is melting an average of **260 billion tons** of ice each year into the seas nearby, this of course resulting in the rise of sea levels world-wide.

The opposite pole is of course affected too, but is often overlooked because of its massive coverage and impressive sunlight reflective ability. The Antarctica cap realized the lowest ever drop in ice coverage in March, 2017. In July of that year – the middle of winter in the southern hemisphere - a **trillion-ton section** of Antarctica's Larsen-C ice shelf broke away from the south polar cap, and as with the northern losses, this attributes to rising of global sea levels – estimated to be up as much as three feet by 2050.

"There's no evidence that anything is recovering here," said *Mark Serreze*, the director of the NSIDC. "What we've seen historically is a downward trend in ice extent in all months. Superimposed on that are the ups and downs of natural variability. We're going to continue to head downward."

For the northern Arctic ice cap, the 2018 minimum was 1.63 million square kilometers (629,000 square miles) LESS ice coverage than throughout the 1981–2010 average ice minimum. NASA scientists *Claire Parkinson* and *Nick DiGirolamo* have calculated that Arctic ice cap has lost roughly 54,000 square kilometers (**21,000 square miles**) of ice for each year since the late 1970s – the equivalent to losing a massive portion of the North Polar Cap the size of Maryland and New Jersey.

Not just for one year….for every year since 1970.

Graphic courtesy
earthobservatory.com / NASA

Figure 2 – Diminishing Ice Coverage with Arctic
Circle

GREENHOUSE GASES and the Effect From the Biological World

There is an even larger problem in terms of the effect on our world, one which could lead to catastrophic changes in only the next few years. This is the release of seabed methane — a potent greenhouse gas — from the continental shelves of the Arctic Ocean.

Yes, this is the same "methane" that the proponents of the "*Green New Deal*" proposed by a few errant free-thinkers in the United States government want to curtail by eliminating all of the cows on Earth. So called *Greenhouse Gases* are nothing new and scientists have known of their effect since 1824 when Joseph Fourier showed mathematically that the Earth's surface would be far colder if it had little or no atmosphere; greenhouse gases such as Carbon Dioxide, methane, water vapor, nitrous oxide and ozone all contribute to the trapping of heat from our sun and somewhat of an equalization of the Earth's atmosphere over the course of seasons.

A BALANCED and steadily maintained proportion of these gases create a natural greenhouse effect which keeps the Earth's climate livable. Without it, the Earth's surface would be an average of about 60 degrees Fahrenheit (33 degrees Celsius) cooler. Levels of greenhouse gases have varied slightly (sometimes dramatically for short periods) over the Earth's history, but they had been fairly constant for the past few thousand years.

This means that the average temperatures throughout all of Earth have also stayed reasonably consistent over that time. But that seems to have changed dramatically as the population of Earth has risen, particularly over the past 160 years.

The Human Factor

For the most part, planet Earth is content with the existence of BIOLOGY. For perhaps over a billion years, this world has sustained and created new remarkably efficient species of living things, from the basic virus to the organized manipulator: MAN.

Max Roser, in *World Population Growth* (Our World in Data, 2017) explains:
"200 years ago there were less than one billion humans living on earth. Today, according to UN calculations there are over 7 billion of us. Recent estimates suggest that today's population size is roughly equivalent to 6.9% of the total number of people ever born. This is the most conspicuous fact about world population growth: for thousands of years, the population grew only slowly but in recent centuries, it has jumped dramatically. Between 1900 and 2000, the increase in world population was three times greater than during the entire previous history of humanity—an increase from 1.5 to 6.1 billion in just 100 years."

Looking even earlier into the explosion of human population reveals that from 10,000 years before present (10,000 BCE) until the advent of the modern era (i.e., the time of the birth of Christ, 0 BCE) there were less than ONE MILLION people living throughout the entire area of the Earth. Not "billion" – million and far less than that in earlier times.

There is a hugely curious aspect to this seemingly out of control growth of the human race on our planet. While the enormous population continues to grow because of increased life expectancy and a substantially lower death rate per capita just in the past century (See Figure 10), the BIRTH RATE worldwide has dropped substantially, somewhat in part by family number restrictions in China alone.

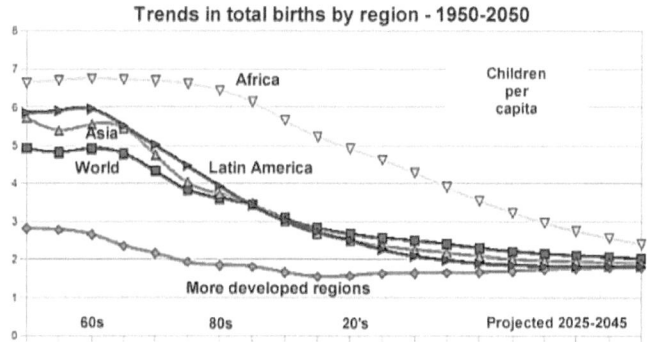

Figure 3 – Birthrate decline worldwide

Nonetheless, as seen in Figure 4, which reflects all aspects of birth, death and life expectancy, the population climb continues in an almost linear path.

By our year 2020, this number will have grown to over EIGHT BILLION – **eight thousand times** the number of humans starting at year 1 A.D (BCE). This suggests that we should look at the effects of industry, directly, as not as destructive to our global balance as the over-population of our planet.

This begs the question: *How many people can this planet support with its limited resources and limited surface area?*

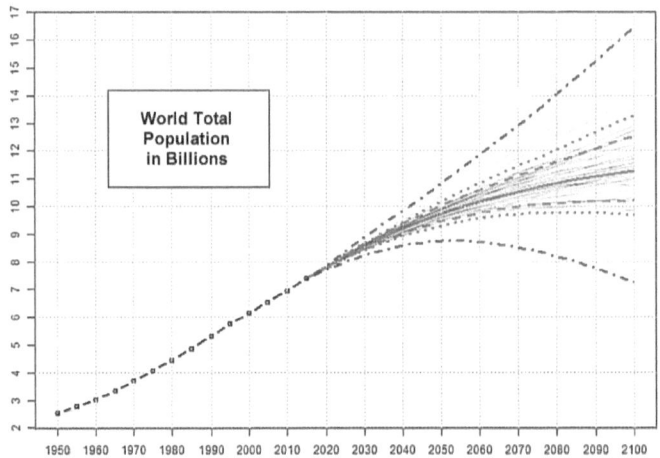

Figure 4 – Actual growth patterns and projections
worldwide
Courtesy UN Dept. of Economic and Social Affairs 2017

A Finite Surface Area of Earth

So far as we understand scientifically, planet Earth is the only world within our space neighborhood that can support human life or life in any complex form. Of that world, only **197 million square miles** is <u>dry land,</u> suitable for habitation. That is only 29 percent of the total area of our planet, the rest being saline-bearing oceans and seas.

Of that 197 million square miles, nearly 33% - one third – is also unsuitable for habitation because of desert conditions, reducing habitable land to **131 million square miles.**

Reduce that space by 10% for all surface areas that are covered in glacial ice, and the habitable land suitable for human use declines to **111.3 million square miles**, still a seemingly large number....until we put all the facts together.

During the dawn of modern man, some 10,000 years ago, it would appear that there were some **111 square miles** for each human on the planet. That seems sufficient and then some.

Advance the civilization clock to today – around 2020 – and with 8 billion humans on the planet, we are left with a shocking **0.14 square mile** per person, or slightly LESS than eight acres, about the space needed to raise about seven cows.

Consider that New York City's Central Park is a whopping 843 acres by comparison.

But you might ask, if the density is so large, WHERE are all these people? The human race has partially solved this diminishing land crisis by living **vertically**, particularly in the largest of all cities such as New York.

These towering human storage facilities can hold a lot of people. Author *Andre Karam*, in quora.com writes:

"It depends on what's on that block [of the city]. If it's a high-rise apartment building, there might be 10,000 people or more. If it's an office building, there might be nobody "living" on the block, even though the office building might have 10,000 occupants working there."

Karam continues his assessment: *"My particular block is in a residential part of Brooklyn and is filled with low-rise apartment buildings. My own building has only 6 apartments, two of which have only a single inhabitant. So my building has fewer than a dozen residents. The entire block has 32 buildings (I just counted), some of which are commercial on the first floor and residential on the next two. So I'm guessing that there are between 500–750 people living on my particular block. But that's just a guess. By comparison, my apartment on Roosevelt Island (13 floors) had 500 units and about 2000 residents….."*

Is "going up" going to solve the problem with over-population? Really?

Note that the growth in human population is something of a remarkable phenomenon. This growth was not and is not linear….it is exponential, like the fission explosion of a hydrogen bomb. I invite you to compare the population growth graphic below (Figure 2) to that of the global temperature rise and note the similar growth during the same period of time.

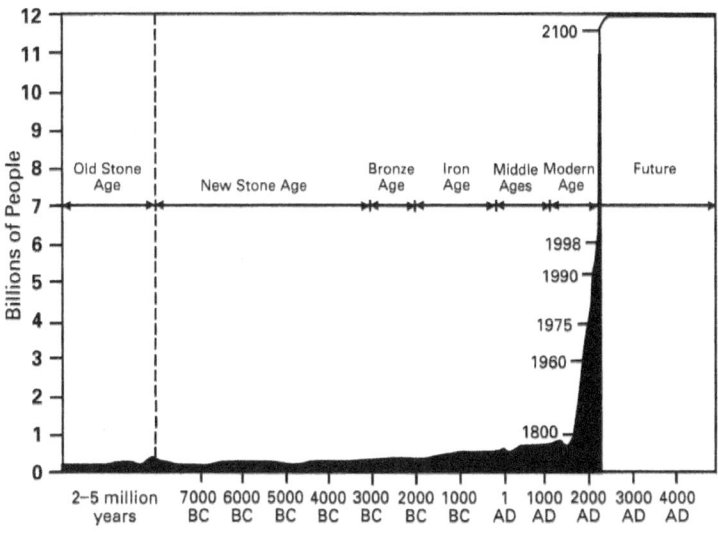

SOURCE: "Population: A Lively Introduction," Joseph A. McFall, Jr., *Population Bulletin*, Volume 46, Number 2, October, 1991, pages 1–43, Population Reference Bureau, Washington, D.C.

Figure 5 – Human Population Explosion Circa
1600 AD

*   *   *

I cannot emphasize enough: the Earth is a finite planet with finite resources. It is incapable of "growing" simply because this biological entity known as "the human race" cannot keep its propagation in check. Eventually, perhaps quickly, this world will run out of room.

The Influence of Natural Factors

Human population and destruction of land area activity are not the only detrimental aspects that affect the Earth's climate. Volcanic eruptions add thousands of tons of ash and water vapor, as well as harmful chemicals into the atmosphere and onto land every day; as the Earth has equalized since its beginning, this aspect has become less and less of a factor, but still is significant in terms of particularly short-term effects.

Cyclic fluctuations in solar radiation from sunspots and solar wind, can occur in very short periods of days, with cycles of 7, 11, and even hundreds of years in period. The Earth's position relative to the sun can also influence global climate change, particularly if solar activity is very high or low which can result in warming and cooling, respectively.

Even large-scale weather patterns such as El Niño and major meteorological changes that might be triggered from solar activity or flux in the Earth's magnetic field might result in fluctuations in temperatures worldwide for years or even decades.

It must be known that small warmings and coolings of the Earth's atmosphere when measured and averaged globally, will occur a single or one time events that can be tagged as such. The volcanoes will emit light-blocking ash that will temporarily hinder the passage of sunlight to the surface, thus resulting in a cooling effect for a short time, maybe as much as several years. Meteorologists are able to pinpoint the effects of El Nino to specific cycles, and even cycles of hundreds of thousands of years (i.e., the Ice Ages) can occur in repeating cycles that are NOT related to the steady climb in Earth's average temperature.

"Even though we're not setting records every year—and we don't expect to because of natural variability—we're not any where close to the averages we saw in the 1980s, 1990s and before," said *Walt Meier*, a senior researcher at the National Snow and Ice Data Center.

That being said, all climate models that are used by climatologists to monitor Earth's temperatures take those natural factors into account. When all of the "natural influences" are combined and considered, these changes in solar radiation as well as particulate matter, steam and noxious gases from volcanic eruption, have contributed only about *two percent* to the recent warming effect. Thus, ninety-eight percent of this rapid warming comes from greenhouse gases and other human factors such as overpopulation and abusive land use over time.

FINITE RESOURCES FOR INFINITE
POPULATION GROWTH

Water, water everywhere....?

It may seem odd, but the fact is that there is the same amount of drinking water on earth as there has always been, only there are more people drinking it.

Figure 6 - Graphic Courtesy US Geological Survey
Clear drops represent ALL potable water on planet
Earth

Freshwater makes up a very small fraction of all water on the planet. Only 0.007% of all water is drinkable. Most of Earth - nearly 70 percent – is covered by water, but only 2.5 percent of it is potable or fresh. The rest is salien/salt infused and ocean-based. Now take into consideration that just **1 percent** of our freshwater is accessible, the rest trapped in glaciers and snowfields. There is not much water available for 8 billion people.

It becomes somewhat distressing, or certainly should, to each of us when we learn of just how little water there is. As stated, the water is not "going anywhere" away from planet Earth, but there are 8 billion people using it instead of the less than 1 million just 2000 years ago.

So where is our water coming from? Certainly not the massive oceans the blanket our planet; that water is salenated - contains dissolved salts that cannot easily or inexpensively be separated from the liquid to make it potable.

Here is a breakdown of the only sources for potable drinking and consumable water that remain; you likely will be stunned by the fractional amounts available.

SOURCES OF POTABLE WATER

Fresh groundwater (may contain contaminants) – 2,530,000 cubic miles
Fresh water lakes – 22,000 cubic miles
Atmosphere (in form of humidity, clouds, suspension) – 3,100 cubic miles
Swamps of Earth – 2,750 cubic miles
Rivers – 510 cubic miles

Surprised how little there is? You should be and you need to be.

How much water does the average person on this planet use or consume per day? Each of us is using more and more water every day according to statistical studies, likely because of carelessness in the manner in which it is used. The average reveals that each person uses about **80-100 gallons** of water per day, the largest usage being that of flushing the toilet, followed by your daily shower or bath.

The inhabitants of the United States are the largest water consumer by far.

Worldwide, considering 8 billion people but eliminating an arbitrary 2 billion who may suffer from famine and drought, this will leave 6 billion humans each of which is using as much as 100 gallons of fresh water daily. Thus, our worldwide average daily consumption of water by all humans is approximately **600 billion** gallons.

One square mile of water contains 1,101,117,147,352 US gallons – about 1.1 trillion gallons. So, roughly estimating, we people of Earth use one square mile of water every TWO days.....about the amount of water that flows over Niagara Falls in a month.

And even knowing this, the average golf courses throughout the world are the top users of water – requiring between 100,000 and one million gallons annually for their daily watering, this according to the Alliance for Water Efficiency.

As we continue to allow the faucet to run while we text on our phones, or brush our teeth, you may want to remember that "water" is really a big deal. In fact: The Biggest.

Nothing on this planet will survive without fresh water, potable water. Many environmentalists and climatologists believe that fresh water will be more valued than gold or diamonds within this century. Perhaps wars will be fought over water rights; if conservation does not serve to provide the needs of a coming 16 billion people, then water rationing may have to serve that need.

I Think that I Shall Never See…..Another Tree?

Contrary to beliefs in your typical classrooms of higher learning, the world is slowly running out of trees. Deforestation is rapidly annihilating the forests that once dominated the landscapes of Earth. Whether for toilet paper or for the house you live in, our disregard for the forests of the world is slowly leading indirectly to Greenhouse gases out of control.

About 300 years ago there were on the order of as many as **10 trillion trees** throughout the world. Even then early culture consumed many of them without regard for replanting, such as the *Cahokians* of the Mississippi Valley did only 1000 years ago. Building great walls, palisades, structures and wooden monuments, those Native Americans soon found that it was no longer a mere walk to the woods to find great towering cedar or oak trees for their projects. Even farther fled all of the game animals that were the food source for their people, as the protective canopy of trees disappeared around them.

And nothing has changed since….except to perhaps get worse.

There are many educators and "experts" who believe they are in the know on the state of the world's forestation situation at this time; they are quick to tell you that replanting of tree farms (pines for rapid harvesting, not hardwoods) has led to more trees today than in any previous time.

They are wrong. Where once were those 10 trillion trees are now merely **THREE trillion** trees, according to global tree surveys done in 2016. Even more alarming is that we are using an astounding 15 billion trees each year for toilet paper, timber, clear cutting, housing and other human requirements.

In the United States alone, there are over 128 million houses and an additional millions of business structures. The number of trees required to build a typical 2600 square foot home is a minimum of 44; the homes now standing in the USA were once over **5 billion trees**.

According to the National Geographic Society, the world lost over **512,000 square miles** of tropical Rain Forest to human deforestation between 1990 and 2016. Most of this lost resource will remain lost forever. Perhaps someone needs to shake the person sitting next to him or her and explain a few very important facts about human survival as it relates to trees.

Let us start with the aesthetics of the trees as they relate to the solitude and sanity of mankind. Only people who have never entered a forest do not know the tranquility that overcomes the human spirit, whether in grief or happiness, when walking into a forested world. Home to thousands of species of life within every acre, there is peace and common ground between Mother Earth and humans within the woods of the planet on which we live.

Those same aesthetics are realized, but in a much different and more practical way, to all of the forest animals and plants that rely on the perfect balance of the woodland to all of the living structures – plants and animals – that live within the forest. It is habitat, protection, survival, food....indeed the forest is life itself for all that live within. Without the trees, there is no life.

Figure 7
Illegal deforestation of 3,050 square miles of the
Brazilian Tropical Rain Forest in 2018
Image courtesy BBC

In the tropical rainforests of South America, illegal deforestation is resulting in what likely will be demise of the most important ecosystem on the planet Earth. All animals on Earth depend on the vital gas OXYGEN for life....it is the breath of all that lives in the animal kingdom. The rainforest is responsible for over **20 percent** of all oxygen that is available on Earth....every molecule of it from the priceless trees.

Not just the rainforest, but all trees of Earth provide balance and life for pretty much all living things that have evolved on the planet. Without trees, there would not be enough oxygen for even a handful of humans to exist here. Each deciduous tree (broad leaf, like oak, maple, mahogany, teak, hickory) can provide over **260 pounds** of oxygen every year. Two of them can supply a *family of four for a year* with this breath of life.

But remember: we are losing 15 billion trees every year; those that we replant are not deciduous trees, but pines with such little photosynthesis capability as to not come close to the human requirements for oxygen on a yearly basis.

Note from the graph on the following page the correlation of population growth to deforestation. Notice the remarkable similarity to the earlier comparisons for global temperature increases and human population of Earth.

Not only do these trees give freely this vital oxygen, but they also assimilate Greenhouse gases that contribute to the ever-increasing warming of our planet.

In all of these factors leading to global warming and climate change, there is one dominant common denominator: uncontrolled   human   population growth.

World population and cumulative deforestation, 1800 to 2010

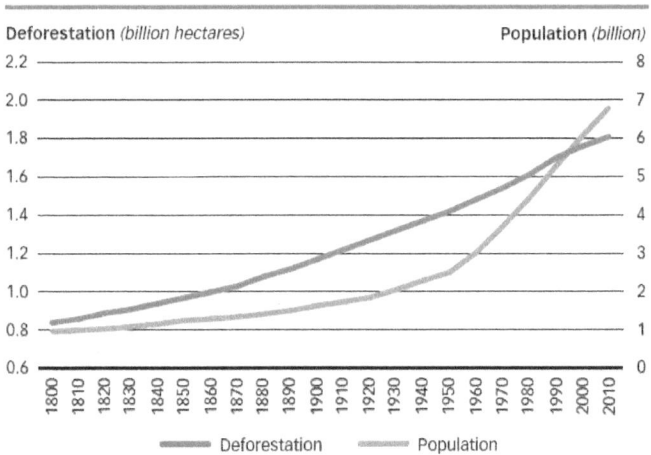

Sources: Williams, 2002; FAO, 2010b; UN, 1999.

Figure 8

One such example of the severity of forest destruction is somewhat of a "catch-22"; a healthy green deciduous tree provides enough oxygen in one year for two people. While those two humans are inhaling and utilizing the oxygen that the tree provided them, they are required to EXHALE a virtual poison to the human body – carbon dioxide.

It just so happens that the tree once again steps in to save the human race; each human exhales about 2.3 pounds of carbon dioxide each day if activity is normal. That is a yearly total of nearly 840 pounds of carbon dioxide – a Greenhouse gas that is responsible for temperature increase of our atmosphere – for every human on Earth.

One healthy tree will utilize 48 pounds of the gas per year, not nearly enough to keep up with the exhaling population of 8 billion humans. For every human who wishes to breathe normally, it is going to require some 23 healthy deciduous trees to rid the atmosphere of his or her carbon dioxide.

Combine that effort with the additional release into the atmosphere of all of the human-related industries and activities from which either carbon dioxide or carbon monoxide are emitted, and the need for massive amounts of healthy forests is immediately realized.

While the present number of trees on earth are sufficiently providing oxygen for our survival, there are rapidly becoming insufficient resources in our rapidly diminishing forests to naturally eliminate the waste of both industry and mankind.

One final – but very important – point to illustrate about this quandary between the expansion of the human race and trees: *infrastructure resources.*

We have already illustrated that the world's population of human beings is spiraling out of control. In this early part of the 21st century it has once again become "popular" and trendy for western civilization families to encourage large families; two seems to no longer be enough children even though there appears to be such little family bonding in this modern age. We are in an era where four to six children are desired and even encouraged from one set of parents to another.

In other portions of the world, particularly the third-world countries and impoverished nations, there is little or no attempt made for birth control; efforts to provide contraceptives to uneducated populations has served little or no purpose, with condoms being used to transport water and worn as gloves.

Figure 9

No human wants to give up his or her life and we do not want nor expect that to happen. Nonetheless, we have reached the critical mass of both resources and space.

Infrastructure Resources and Population

Consider this interesting paradox that nearly all humans are contributing toward daily.

Every human has the ability for logic – we are the only creature on planet Earth that can reason and rationalize. We can figure out ways to survive, we can create works of art, design great buildings and bridges, expand our frontiers simply through ingenuity. Nothing else living can do that, only us. A raccoon cannot, although he can figure out how to pop the lid on any garbage can. The rest of the living world lack the human ability to conceive concepts and develop ideas into reality.

Also, while on the subject of the raccoon we should note that the raccoon is not destructive to the environment, whereas the human is. Nor are birds destructive, snakes, bugs, spiders, elephants....nothing, but humans.

We are the only creature – and there are now 8 billion of us and growing – who can alter his environment to make it what he wants it to be. The only creature to intentionally create waste products and then carelessly throw them along a roadside. Over the countless years, all other biology actually contribute to the success of the world; they give back, even with the deaths of their bodies which decay into soil for future living things. Humans do not even do that: we cremate or embalm our dead bodies so that they are worthless flesh that enriches nothing of the world.

Even in death, the human is a selfish creature.

So let us examine the "human drive" and link it to the chain of Greenhouse gases and the rapid deterioration of our planet that is leading to the melting of the polar ice caps. Following is an exercise to understand the human desire to "keep up with the Jones'" and this drive appears implanted in all of us, myself included.

Babies are born and rapidly grow to maturity. Along the way, all traits of human rationalization begin to develop and the influence of society and peers begins to take an equally dominant role in thought processes.

Every one of the children grow up and if possible one of the first things we all want is an automobile or means of transport, usually requiring polluting fossil fuel propulsion. Soon, the human finds more mature needs, perhaps a home for a growing family, and seeks a place to put both the automobile and the home. To make place among the many other homes of other humans require "neighborhoods" and "cities" and these developments require that growing living green things – many times deciduous trees – be cut down to make room for each domicile in the neighborhood.

So we at some point, let us say, have a family of four now living in a small parcel of land where trees once stood producing oxygen and purging our air of poisonous Greenhouse gases.

The people within the household, as well as all neighbors, of course want a place to park the vehicles that they own when they are not out burning fossil fuels in them and emitting noxious Greenhouse gases. So green grass and plants (also capable of photosynthesis to produce oxygen and reduce poisons) are scraped away and cement is poured in its place.

But that is not enough: there must be connections throughout the land areas where the vehicles can rapidly traverse from one destination to another – highways and streets, all requiring the removal of green growing things and replacing that with carbon dioxide and Greenhouse gases as well.

Shops must be built, schools constructed, stadiums erected, bridges for crossing those pesky water sources....more concrete, less trees. And, yes, we must have huge parking lots made of asphalt or cement on which to park our automobiles at these shops, schools and more.

Aside from the marvelous process of photosynthesis, green plants have a remarkable capability to cool down the immediate environment in which they reside. Things are simply cooler in the woods than they are in a parking lot. *But stop*! We are reaching the point where there are more parking lots, houses, roads, buildings and bridges than there are forests.

I invite you to drive through any large metropolitan area on an Interstate Highway in the middle of the night on your next trip from one destination to the other. Watch your automobile thermometer – most are quite accurate. As you approach the large city, take a look at the night time temperature; perhaps it may read "71 degrees"; but keep watching. As you pass through the most dense part of any metropolitan area, that temperature will rise usually at least THREE degrees, all of that being heat which was retained in the cement and steel during the hottest part of the day.

While we are on the subject of roads and their influence on the diminishing forests and the ever-increasing global temperatures, join me in a quick look at the resources that the earth provided to construct the Interstate Highway system. These are resources for which trees were bulldozed to dig the earth, resources which required millions of years for the Earth to compose, and resources that we will never replenish in the future of mankind.

As of the first decade of the 21st century, there were 97,000 miles of Interstate Highway traversing just the United States. Those highways, made of cement on treeless avenues hold over 55,000 bridges of cement and steel. The stone used simply for the roadways is enough to build 700 more pyramids in Egypt. The cement poured on top of that roadway is enough to have constructed SIX sidewalks to the moon – 1.4 million miles.

And every inch of this required the removal of an Earth-saving green growing thing…but we can get from that one destination to another very quickly.

You see, the human specie is perhaps the most selfish of all in all respects other than survival. Yet humans have the capability through logic and creativity to ALTER the environment around them to survive, or at least survive more easily.

One more example that demonstrates the importance of infrastructure over the future of the climate of the Earth, that being San Francisco's Golden Gate Bridge: A marvelous feat of engineering and design, but one wonders if it really had to be built. Is there not a long way around getting to Oakland?

The Golden Gate Bridge is comprised of a remarkable 80,000 miles of steel girders, cable, rebar, bolts, nuts and other gear. Every ounce of that material was produced by Mother Earth long before mankind ever walked anywhere near the oceans. Even with the great bridge completed, worldwide we still continue to pillage the Earth of 43 million tons of iron ore annually, all from strip mines where once stood the oxygen-rich, poison-devouring great trees of Earth.

What becomes of this Earth if we continually drain it of its components and alter it in a way that only mankind can do?

Reaching the Point of No Return

Early in the 1970's I traveled throughout the United States and lectured on the dangers of uncontrolled population and the limited resources available on this finite planet Earth. Perhaps I was a Pioneer in that respect, but they always say that you can tell the "...true pioneers by the number of arrows sticking out of his back."

And that is the way it was.

In 1972 people simply did not give any mind to climate change; we had Vietnam, acid rock, hippies and the beginning of political posturing.

From then until today I emphasize that the root condition that leads to the changing of any world in the scale that we are seeing today is NOT a natural phenomenon like a solar cycle or posturing of the magnetic field of this planet. The root cause is very simple:
OVER-POPULATION.

The solution to the problem is two-fold:
  1) fix the population problem
  2) find a way to fix a broken planet.

Neither of which is at all simple and frankly neither of these parts of the solution is likely ever going to take place. It seems that we have stretched our finite Earth beyond her abilities to support and maintain our lifestyles. The human race has outlined, staged, and produced perhaps the death sentence of all biology.

My efforts in the 1970's and those of others are not unlike those of today; each generation of humans believes firmly that the issue does NOT involve them and that somehow this entire idea of the human race driving itself some what *purposely* to extinction is abstract and absurd.

Consider that the problem has become even more pronounced in the 50 years since the 1970's through the advent and progress of modern medicine. Quite simply humans are living longer; since just 1970 the life expectancy has risen from 67.1 years (male and female combined) to an incredible 82 years for both male and female by 2018 (source: Berkeley Medical).

This becomes even more of a factor when you learn that life expectancy in 1500 Europe was only 35 years of age.

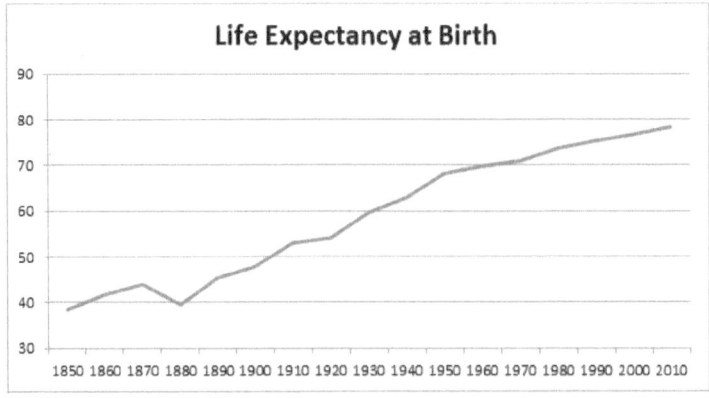

Figure 12 – Life Expectancy in Years 1858-2010+

Obviously the human race is seeing a longer life span with fewer deaths. All the while, in relation to the total birth rate which has dropped significantly over the past century per capita, the total population still surges higher and higher with all aspects considered:

- total decline in birth rate per 1000 humans
- total increase in life span of average human
- significant decrease in annual death rate

The sad fact is: the decline in the birth rate at present per capita simply cannot offset the population surge and thus the overall occupancy of the planet continues to rise (Figure 4) and the planetary occupancy by the human race continues to grow at a linear rate.

If the human race suddenly today ceased just where we are, then biology would recover; it always does once left alone. But that is not likely to happen and the fantastically silly overtures of "New Green Deal" and "Going Green" is juvenile and ineffective at best. It may simply be too late.

The fact is: life existed on planet Earth long before the appearance of mankind and sustained just fine without altering the planet to do so. Unless the only creature which CAN detrimentally alter his environment is the human, if our specie disappears and leaves just any aspect of biology intact before doing so, biological species that do not have the ability to reason and logically create will once again thrive on the planet.....depending on what the human race leaves behind.

An Unpleasant Prognostication

The British scientist/submariner Peter Wadhams perhaps first provided illustrative proof of just how quickly climate change was altering the northern Arctic Circle. The rapid seasonal decline of the ice caps into the fall are telling, but the summers have really provided scientists with evidence that we may be losing the North Polar Cap ice, and this loss may be permanent. As early as 2030 the Arctic Ocean could lose essentially all of its ice during the warmest months of the year—a radical transformation that would destroy virtually all of the Arctic ecosystems and disrupt or destroy many northern communities, if not all communities along the coastal areas of Earth.

Throughout its history, we know that the Earth has worked though cyclical climatic changes, some of which happen at fair regularity. The last of these was our last **Ice Age** when the difference between average global temperatures today and during those ice ages was only about 9 degrees Fahrenheit; the swings have tended to happen slowly, over hundreds of thousands of years. We are aware of these and we know that they have absolutely nothing to do with anything that the human race might be doing on our planet.

But now concentrations of Greenhouse gases are rising dramatically – because of mankind's industry as well as human overpopulation leading to the destruction of the cycle of photosynthesis. Earth's remaining ice sheets over Greenland and Antarctica are shedding their glaciers at an alarming rate that increases with every passing day. That extra water could raise sea levels significantly, and quickly. By 2050, these melting glaciers will provide enough volume in the oceans and seas to raise the sea level by as much as 2 or more feet.

The cycle does not manifest itself just in melting ice caps and homeless polar bears. The rising temperatures will ultimate lead to *very extreme weather*, something that meteorologists are already seeing yearly in short term patterns: volatile storms instead of gentle rainfalls, increases in cyclonic activity from rapid upthrusts of heated air in the form of tornadoes and hurricanes, unpredicted droughts where none should be.

Indeed, the entire climate of planet Earth may shift creating deserts where none were before, swamps amidst the sands of Death Valley. With such meteorological changes come obvious side effects in the form of shorter growing seasons for crops for our 8 billion people, limited grazing and habitat areas where wildlife can thrive and propagate.

And ultimately, we are facing the possibility of the loss of precious water, the one substance that is required by all of biology to survive. The Earth can and will survive without potable water; the world of biology cannot. Just from the melting of the polar glaciers into fresh water, we have lost millions of cubic feet of water already as the pure essence falls diluted into a salty sea.

Has the cycle reached its "critical mass" and now unable to be reversed? Will popular social efforts such as "Going Green" help in any way whatsoever at this point in a global evolutionary crisis? In only a few – perhaps two – generations of the human race might we know the answers to whether the human race will have a planet capable of sustaining life without ever leaving this world.

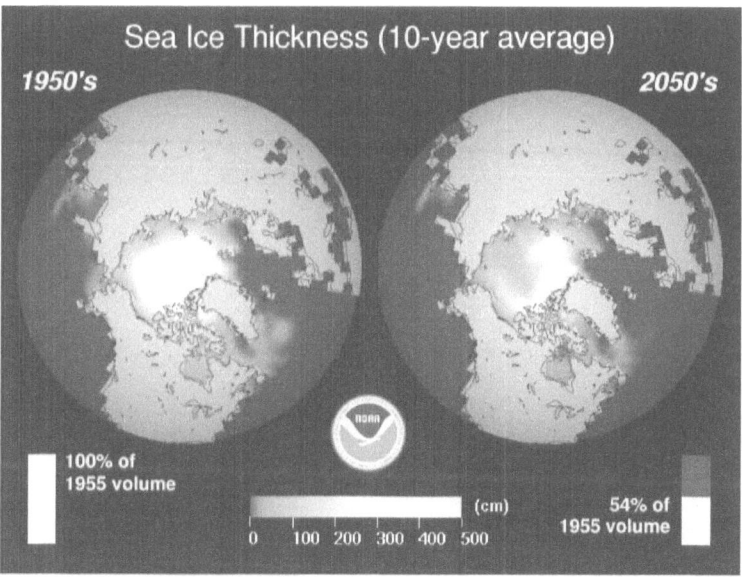

Figure 13
North Polar Sea Ice Thickness Decline –Century
Courtesy NOAA

Other Publications by P. Clay Sherrod

Available by contacting ASO at **www.arksky** or direct
through **drclay@tcworks.net**

Also available through *Amazon.com* or *Barnes & Noble*

Sherrod, P. Clay, *The Arkansas Sky Observatorie's
Supernova Search Atlas and Guide*. A study guide for
the discovery of extragalactic supernovae. 2018, 164pp.
Large format soft bound. 350 galaxy photographic atlas.
$19.95

Sherrod, P. Clay, *The Genesis Project: Creation and
Science through Art* / Premium Collectors Large Format
Edition / Hardbound coffetable edition full color – LuLu
Publishing. Original artwork and philosophical
discussions of Biblical Genesis by author   Arkansas
Sky Observatories. $95

Sherrod, P. Clay, *The Genesis Project: Creation and
Science through Art* – Arkansas Sky Observatories –
Premium hardback edition, full color - Original artwork
and philosophical discussions of Biblical Genesis by
author. 68 full color pages.   Arkansas Sky
Observatories. $34.95 (hard)

Sherrod, P. Clay, *A Complete Manual of Amateur
Astronomy,* Simon and Schuster, New York. 1982.
Hardbound, softbound.

Sherrod, P. Clay, *The Constellations: Tours for
Computerized Telescopes Vol. I* 2017, Lulu Press,
Study Guide packed with illustrations and photos; 396pp
- Premium paperback - $19.00

Sherrod, P. Clay, *The Constellations: Tours for
Computerized Telescopes Vol. II* 2017, Lulu Press,
Study Guide packed with illustrations and photos;
550pp - Premium paperback $29.00

Sherrod, P. Clay, Telescope Rx. 2017, Lulu Press, 1st ed, 352pp Large format study and self-help guide; Premium paperback ed., $29.50

Sherrod, P. Clay , *Catalog of Cometary Orbits, Arkansas Sky Observatories* - 2017 , large format, Premium Paperback edition - 296pp. $29.95

Sherrod, P. Clay, *Footprints of Fallen Giants – Pathways to Extinction* / (Dinosaur evolution philosophy and discussion) large format publication of original manuscript with notations, Premium Paperback edition with many original illustrations throughout – 314pp 2016, Lulu Press - $29.95

Sherrod, P. Clay and John Berky – *In Vitro Toxicity Testing: For Carcinogenesis, Mutagenesis and Toxicity* – Franklin Institute Press for the US Food and Drug Admin. 1977, 433pp., paperbound. Available on auction sites and through Amazon. $120+, current market price

Sherrod, P. Clay, *Motifs of Ancient Man*, Rock art and pictographs of Arkansas. University of Arkansas Press, 1983. Some copies available via Amazon and rare book sellers. Paperbound, 160pp. Illustrated. $125+ current market price

Sherrod, P. Clay, and M. Rollingson, *Surveyors of the Prehistoric Mississippi Valley* – *The Toltec Module*. The incredible engineering of the ancient mound builders revealed. 1987, Arkansas Archeological Survey Research Series #28. 162pp. Paperbound – Available through U of A, Amazon, Toltec Mounds and Cahokia Mounds State Parks

Sherrod, P. Clay, *Rocks of Prehistory: The Gumlog Creek Rockshelter* – Premium paperback edition, many illustrations and discussions of prehistoric occupation and anthropological overview of regional Native Americans. 112pp. Lulu Press, 2016 $18.00

Sherrod, P. Clay, *Ancient Quests of the Double-Peaked Mountain: The Gunbarrel Petroglyph* – A captivating discussion highlighted with fascinating illustrations throughout. 72pp. Premium paperback edition. Lulu Press - $15

Sherrod, P. Clay, *You Twit Face: Your Complete Guide to Nomophobia* (our obsessions with the "Smart Phone") – Premium Paperback edition - Hardback edition, 234pp $29,95; soft $19.50 Lulu Press, 2016

Sherrod, P. Clay, *Life Lessons from the Naked Boy*. 2016, Lulu Press, Hardcover edition , Premium Paperback edition - 266pp - $19.95 soft; $29.95 hard cover

Sherrod, P. Clay, *The World's Last Peanut – a Curious collection of politically incorrect opinions and thoughts*. 2018, Arkansas Sky Observatories Publications. Softound, 130pp. Lulu Press.

Sherrod, P. Clay, *Cookin' In The Woods*. 2018, Arkansas Sky Observatories Publications. Spiral Bound cookbook, 86pp. soft cover, $14.

Sherrod, P. Clay, *Fireflies From Orion: The Poetry of P. Clay Sherrod* . 2019, Arkansas Sky Observatories, softbound, 349 pp.

Sherrod, P. Clay, *Fission Population* . 2019, Arkansas Sky Observatories, softbound, 37 pp.

AND MORE.....please see the Arkansas Sky Observatories' Publication Tab at: http://arksky.org/publications

www.ingramcontent.com/pod-product-compliance
Lightning Source LLC
Chambersburg PA
CBHW021044180526
45163CB00005B/2284